PORTRAIT SERIES　　　　　　　　　　　　　　　　　　　　　　　　　　　　　　　　PS 12

The Stratford-upon-Avon and Midland Junction Railway
a pictorial survey by Stephen Thompson

Byfield Station, taken from the overbridge.

THE OAKWOOD PRESS

Photographs and text © 2019 SMJ Society
The Oakwood Press
54-58 Mill Square, Catrine, KA5 6RD
01290 551122
www.stenlake.co.uk

ISBN 978-0-85361-461-6

Printed by
P2D Books, 1 Newlands Rd, Westoning, Bedford MK45 5LD

Publisher's Note

The photographs in this book were taken by Michael Clarke's uncle, Stephen Thompson, a part-time commercial artist who had a passion for railways and owned hundreds of railway books covering lines all over the world. In August 1961 he set out, in his E-type Jaguar, from his home in Northampton and compiled a photographic survey of the Stratford-upon-Avon & Midland Junction Railway. He set the photographs, with typewritten captions for each one, on pages like the one reproduced on the right. After Stephen's death the notebooks passed to his nephew Michael Clarke, who in turn gifted them to the SMJ Society. It was the SMJ Society who approached us to publish this book.

This book is a selection of the photographs and text from the notebooks. The majority of the pictures are included, except where there are two or more very similar images. Stephen's introduction to the railway and some other passages in the notebooks were taken verbatim from OL10 *The Stratford-upon-Avon & Midland Junction Railway* by J.M.Dunn, Oakwood Press, 1952, and these are reproduced in this book. However, the majority of the text has been taken from his notes explaining the location of the photograph. The text is more or less as he wrote it in the early 1960s, with some editing because the book's layout is different to the pages. In 1961 when the text was written sections of the railway were open or had only recently closed and it reflects that moment of time and Stephen's observations on what he was photographing.

Acknowledgements

Since 2009, the SMJ website has grown in both membership and contents. Early members whose contributions helped set the standard for the site were Dick Bodily, Lloyd Penfold, John Jennings, Robin Cullup, Si Donal, Alwyn Sparrow and the Great Western Railway author, Adrian Vaughan.

Gentlemen ... and those not named ... but you will know who you are - "Thank you"

A special note of thanks for the generosity of Mike Clarke, who kindly gifted the volume and manuscript to us, in the hope that it would be useful in adding to the history of the last days of this particular railway heritage.

Our hope is that it will prove useful in modelling and other studies related to the railway.

And a special note of thanks to Richard Stenlake owner of Stenlake Publishing and Oakwood Press, for taking on this project.

Andy Thompson & Peter S Lewis

Contents

Foreword	3
The Stratford-upon-Avon & Midland Junction Railway Website	4
Return to the Stratford-upon-Avon & Midland Junction Railway	4
Introduction	6
The History of a Line	6
ROLLING STOCK of the Stratford-upon-Avon & Midland Junction Railway	8
Component COMPANIES of the Stratford-upon-Avon & Midland Junction Railway	9
Easton Neston Mineral & Towcester, Roade & Olney Junction Railway	11
Northampton & Banbury Junction Railway	24
The East & West Junction Railway	42
The Edge Hill Light Railway 1920 – 1925	60
The East & West Junction Railway (continued)	63
The Stratford & Moreton Tramway	82
Evesham, Redditch & Stratford-upon-Avon Junction Railway	85

Foreword

There are many nooks and crannies of closed railway lines which never quite get the attention they deserve. The Stratford Upon Avon and Midland Junction Railway was always a bit of a basket case, never made significant money and was never well-used, but it still is an important part of our railway heritage as these photographs show.

It was one of those lines which, had there ever been a strategic body to determine where railways should be built, would probably have never seen the light of day. However, that does not mean it was useless. It gets a mention in *Fire and Steam*, my history of Britain's railways because it did good service in the First World War. As I quote, from Adrian Vaughan's excellent *Railways, Politics & Money*, 'a heavy traffic in iron ore, iron and steel ran over this serpentine byway to and from south Wales and the Midlands'. Admittedly, as Vaughan emphasises, the track on a railway whose cruel acronymic nickname was the 'Slow, Moulding & Jolting' was never in good shape and was 'completely unsuitable' for such dense traffic. He was, though, impressed by the 'high standard of enginemanship' required to operate on the 'ramshackle' track.

This just goes to show how even the most unlikely parts of the rail network could, at times, assume great importance. That is why it is so important to ensure that their histories continue to be available and studied for future generations.

Christian Wolmar, May 2017

The Stratford-upon-Avon & Midland Junction Railway Website

In the beginning...

Many years ago I went to visit my brother Toby who lived in Towcester in Northamptonshire. I would like to add at this point I have always been into steam railways with a bias towards railway history. I live in Devon, a county where many of the railway lines are just that; history, so old trackbeds, street names and former railway structures litter the Devon countryside.

With that in mind, my brother's wife; Sarah asked if we could pop to Tesco for some milk. And so the story started! Off we went, my brother and I, and as we approached Tesco we noticed an old bridge, railway I thought! We entered the shop and my brother casually said "of course you know this is where the station was!?" And that was it, I was bitten. On our return we switched the computer on and headed for Google Earth. Sure enough the images of the Towcester area showed lines of hedges radiating from Towcester on both sides! Two to the east ran parallel; there was mention of quarrying. A trip out in the car was called for, so with camera in hand we set off! Sure enough on both sides of Towcester we found railway cuttings, old bridges and structures.

At the time I worked as an IT teacher and was familiar with the making of websites so on my return home I thought a website featuring the railway history of Towcester was called for, featuring all those photos I'd taken, but what to call it? www.towcestersrailwayhistory.co.uk was born! And from the setting up of the SMJ site, came Steam Tube (with Peter S Lewis), so I was able to rustle up some interest from the membership, some of which was based in the Northamptonshire area. Things gathered momentum, more information was found and the website(s) grew.

Nearly twelve years later we have the website you see now at www.smj.me I soon discovered that this wasn't just about the railway history of Towcester, but about a history of a long gone railway called the Stratford Upon Avon & Midland Junction Railway, more affectionately known as the SMJ. I have since been to and recorded every station site and junction along the line, written articles for magazines, lectured on the subject, including at Oxford, and grown the website into a complete and full history of the SMJ in photos, articles and memories from those now gone who remember the line.

The SMJ was built on the premise of huge amounts of iron ore being available in the Blisworth area. An east/west line might connect this iron ore with the blast furnaces in South Wales. The railway mania of the time enabled the line's construction to start but it never was a success and despite grand plans and fool-hardy extensions; passengers numbers never grew. Beeching was sharpening his knife, but in the end it was not needed. By the mid to late 60s the tracks were being lifted. Another much-loved but little used line had gone.

Andy Thompson
andy@smj.me
www.smj.me

Return to the Stratford-upon-Avon & Midland Junction Railway

Setting the Scene

About twenty years ago, during a period of convalescence from a work-related injury, I determined to use the time profitably by endeavouring to acquaint myself with the SMJ, since I live at the Broom Junction end of the line (in Bidford-on-Avon)…

I recall that I was just beginning to get the hang of the internet, and Google in particular. So every day I would type in *Stratford-upon-Avon & Midland Junction Railway*, and invariably get the same references scattered throughout cyberspace. I purchased the Oakwood Press booklet which outlines the key historical facts, and came by a copy of Arthur Jordan's book, *The Shakespeare Route*, all with a view to writing my own account of the line, from an observer's standpoint…

I made requests for information via the local paper, and was invited onto a BBC Coventry and Warwickshire Radio afternoon programme to be quizzed by the presenter….

And a few pieces of information came my way, most notably a several page handwritten note from a Mrs Wheeler in Stourport on Severn, whose father had been a driver on the SMJ. She enthused about the "Slow, Moulding and Jolting" thus … "It was a lovely line. At some points you could get off the train whilst it was moving, pick moon daisies at the side of the track, and get back on the train!" What a delightful pastoral image that brings to mind!

Then, it happened! Googling for the SMJ, and at the top of the list …www.smj.me ! So, rather than simply book marking the site, I joined! I thought that it would be a useful resource for my book project… And it was interesting to know how this site had come to birth.

Andy Thompson, of Feniton, Devon, was visiting his brother in Towcester. And whilst they were at the local supermarket buying milk, his brother volunteered the comment that "..this used to be the site of a railway station..".. The rest, as they say, is history.

Not long afterward, I congratulated Andy on the site, its layout and design, and said "If I knew how to do it, I'd set up Steam Tube"…and be able to set a respectful and courteous tone , so evidently missing from YouTube and some other railway-specific websites. The same day, he went to the trouble of setting up a Home Page for Steam Tube – to which we audaciously added the tagline.."The Home of Steam on the Net"!

Both sites have continued to grow..and we hope, have become an early port of call to anyone researching SMJ history and/or steam locomotive activity past and present.

In developing the SMJ site, and Steam Tube, we have been able to interact with some respected voices in railway literature. Adrian Vaughan, best known for his many books on GWR history, and IK Brunel (incidentally, *Railwaymen, Politics and Money* is as definitive a book on the history of the UK's railways from 1800 to 1948 as I have come across); and Christian Wolmar, whose support, particularly of Steam Tube, and his contributions to our monthly magazine *On Shed*, and his willingness to allow us to film him presenting a couple of his book presentation slideshows. *Engines of War*, filmed near to King's Cross Station , and *To The Edge Of The*

World – History of the Trans Siberian Railway, filmed at the Hay Festival. Gentlemen, thank you.

Our position as guardians of the SMJ's history – if that's not too grand a remark to make – led to one gentleman presenting us with the unpublished manuscript and 150 original black and white photographs of the SMJ, when the track was in situ around 1961. This manuscript is the basis of Act 2.

Meanwhile, let us proceed to Act 1, and recount the highlights of the early days, of what was, for a short time anyway, The Stratford-upon-Avon and Midland Junction Railway.

Act 1
Scene 1

Stratford-upon-Avon is a town in Warwickshire, England, that just happens to be the birthplace of the most pre-eminent of all writers and playwrights, William Shakespeare. (23rd April 1564) From a small market town …described in Moncrieff's 1924 guide book *Excursion* as "small, but handsome and airy", and earning its prosperity from commercial traffic on the river from which it takes its name..and with a population in the late 18th century of around 3,000.. it has increased to around 30,000 (120,000 plus in the District) at the beginning of the 21st century..and shows no sign of slowing down, with increasingly large developments on the edge of town in all directions.

On top of this, the town has become one of the top tourist attractions in the UK and for this David Garrick can take the credit, having inaugurated the first Shakespeare Festival celebratory procession to honour the Bard (in 1769). And how do the majority of visitors reach Stratford-upon-Avon? A recent statistic has less than 10% of the estimated nearly 5 million visitors arriving by train. Stratford–upon-Avon is effectively at the end of a branch line, and it could have been cut off altogether from the mainline network had it not been for the sterling efforts of those involved in keeping the North Warwickshire Line open, and which today, in the summer months mainly, sees regular steam train excursions to the town.

Well, it could all have been so different! If only local man William James (Henley in Arden seven miles north of Stratford-upon-Avon) had been able to see through his 1820-21 visionary plans of a line from Stratford-upon-Avon to Paddington – a Central Junction Railway – or if his projected survey for the Liverpool–Manchester line, or the three surveys for the Canterbury–Whitstable Railway in Kent had been enacted… if only… even the Stratford to Moreton Tramway – authorized soon after the Stockton & Darlington Railway – was reduced to a horse drawn affair, when the use of Stephenson locomotives might have produced a different outcome… If only… then Stratford-upon-Avon could have been at the centre of railway history, and been at the top table, not just having to be satisfied with the crumbs!

On 23rd June 1864 an Act of Parliament authorized the making of "a Railway from the Northampton & Junction Railway to the Great Western Railway at Stratford-on-Avon" The wording of the Act is worthy of note, if only to illustrate the extreme confidence that the project started with…" Whereas the making of a Railway from the authorised line of the Northampton and Banbury Junction Railway near Towcester in the County of Northampton to the Great Western Railway in the parish of Old Stratford in the County of Warwick will be of public and local advantage; and whereas… etc…"

Of course, the motive for the line – taking advantage of the iron ore in Northamptonshire and getting it down to the smelting plants in South Wales – would have been commended for its insight, but for the fact that when the line got going, cheaper iron ore was being imported!

Peter S. Lewis

Grid References

The grid reference numbers **(818 535)** in the book were taken from Ordnance Survey 1" to the mile maps 144, 145 and 146 in grid block 42. The modern grid is slightly different to the one used in the early 1960s, however, the references will still indicate roughly the right location on any modern map that uses the National Grid, although the grid block is now coded SP.

Introduction

Travellers from London to the Midlands and the North sometimes notice a single line of railway leading away from the main lines about 60 or 70 miles from London, at Blisworth or Woodford or Fenny Compton. and perhaps wonder where it goes to. This was the Stratford-upon-Avon & Midland Junction Railway, a little known and little used undertaking that traversed some of the least frequented parts of central England. Its history, in places as elusive as the line itself, is a curious example of the survival of the unfit. Traffic obstinately refused to flow east and west along its single track; the main lines disdained, after some cautious experiments, to use the connections it afforded; and yet, in spite of financial straits for which 'embarrassment' would be too mild a word, it survived. That fact alone made it noteworthy to the railway historian; so, too, did the remarkable collection of locomotives that at different times dragged its vehicles over the Warwickshire plain.

The SMJ grew out of two proposals for railway routes to convey iron ore from Northamptonshire to South Wales. Local ironstone in South Wales was scarce and generally poor, and the ironworks had to get haematite ore by sea, notably from Furness. In the early 1860s there seemed a reasonable prospect that the lower-grade ores of Northamptonshire might be profitably transported by rail to the Welsh furnaces; but cargoes of Spanish ore were already beginning to arrive, and throughout the rest of the century South Wales was supplied by them. Only in the First World War, and again for a short time during and after the Second World War, did the railways begin to be appreciably employed for their original purpose. That is the uneconomic background of their unprosperous history.

The History of a Line

This is the story of a railway, a line of no particular merit or distinction. Its only claim to any fame is the appalling financial muddle it succeeded in getting itself into, in the very early days. A muddle so profound, in fact, that when selling seemed the only course open, this was found to be impossible because no-one could discover what belonged to who.

Out of this chaos of the original companies was created the Stratford-upon-Avon & Midland Junction Railway, a line which has sections still in operation today, a continuous process since 1866.

It has never been a prosperous concern, and the very fact that it has continued to function all these years is in itself a remarkable fact.

Some parts, of course, have decayed, such as the Northampton & Banbury Junction, which is now nothing more than a grassy track, or the Evesham, Redditch & Stratford-upon-Avon Junction Railway, which will suffer the same fate; nevertheless the main part continues to serve a useful purpose.

Former southbound spur to the Great Central Railway at Woodford Halse, with the East & West Junction Railway heading off to the right. The spur has been disconnected and now forms a set of sidings.

THE HISTORY OF THE LINE

Evesham, Redditch & Stratford-upon-Avon Junction Railway Line facing west from Wasen overbridge, near Binton Station.

Easton Neston Mineral & Towcester, Roade & Olney Junction Railway at Ravenstone Wood Junction, with the old Bedford-Northampton line on the right of the picture.

ROLLING STOCK
of the Stratford-upon-Avon & Midland Junction Railway

The rolling stock, like the locomotives of the three lines, amalgamated, into-the SMJR in 1909. All belonged to the E&WJR. The oldest and most interesting vehicle was an ancient four-wheeled coach dating from 1850, which had been purchased from the LNWR in the 1870s. It was in regular use as the official inspection saloon until 1909, after which it stood for some time in a siding at Stratford. Later the body was removed from the underframe and placed on the down platform, where it remained until the LMS absorbed the line. It was fitted with the Westinghouse brake probably the oldest vehicle to have one of the continuous power brakes.

The remaining passenger rolling stock consisted of four- and six-wheeled coaches, the latter built by the Birmingham Railway Carriage & Wagon Co. In 1909 the railway owned only four composites, four thirds and eight other coaching vehicles. In 1910, in addition to completion of the conversion of the existing four-wheelers to six-wheelers by insertion of an additional axle, four carriages, (two of which were bogie coaches) were bought from the Midland Railway. By 1916 the coaching stock consisted of 28 vehicles, including ten horseboxes and a hound van.

All were fitted with Westinghouse brake and most with vacuum brake also. At the time of amalgamation in 1909, the coaching stock was painted lake below the waist with cream waist and upper panels. For a short time the upper panels were also painted lake, the waist panels remaining cream; finally, from 1910 onwards, the carriages were painted lake all over, conforming with the Midland style. A list of 1916 shows 171 goods and service vehicles.

Northampton & Banbury Junction Railway. The line soon after leaving Blisworth, on the section remaining in use.

LOCOMOTIVES OF THE S.M.J. RAILWAY

E. & W.J. or S.M.J. No.	*Builder*	*Date built*	*Original owner*	*Type*	*Cylinders*	*Dia. of Coupled wheels*		*Disposal* S = *Scrapped* W = *Withdrawn*
					in. *in.*	*ft.*	*in.*	
1	Manning, Wardle (178)	1866	Crampton	0–6–0ST	11 × 18	3	2	Sold 1910—S. & M.R.
1	Beyer, Peacock (1235)	1873	E. & W.J.	0–6–0	16 × 24	4	7	Sold 1875—L. & Y.R.; S 1899
2	Beyer, Peacock (1236)	1873	E. & W.J.	0–6–0	16 × 24	4	7	Sold 1875—L. & Y.R.; S 1899
3	Beyer, Peacock (1237)	1873	E. & W.J.	0–6–0	16 × 24	4	7	Sold 1875—L. & Y.R.; S 1899
4	Beyer, Peacock (1238)	1873	E. & W.J.	2–4–0T	15 × 20	5	0	Sold 1875—L. & Y.R.; S 1909
5	Beyer, Peacock (1239)	1873	E. & W.J.	2–4–0T	15 × 20	5	0	Sold 1875—L. & Y.R.; S 1921
6	Beyer, Peacock (1240)	1873	E. & W.J.	2–4–0T	15 × 20	5	0	Sold 1875—L. & Y.R.; S 1902
4A			A French Railway	2–4–0		5	6	S about 1880
5A			A French Railway	0–6–0				Sold 1885—Cardiff R.
1	Yorkshire Eng. Co.	1873	A Mexican Railway	0–6–6–0	16 × 20	3	6	
2	Yorkshire Eng. Co.		A Mexican Railway	0–4–4T				
1	Beyer, Peacock (1830)	1879	E. & W.J.	0–6–0ST	16 × 22	4	0	1890—Rother Vale Colliery
2	Beyer, Peacock (1919)	1880	E. & W.J.	0–6–0	17 × 24	4	6½	L.M.S. 2300; W 1926
3	Beyer, Peacock (2049)	1881	E. & W.J.	0–6–0	17 × 24	5	0½	L.M.S. 2301; W 1924
4	Beyer, Peacock (2626)	1885	E. & W.J.	0–6–0	17 × 24	5	0½	L.M.S. 2302/2397; W 1929
5	Beyer, Peacock (2466)	1885	E. & W.J.	2–4–0T	17 × 24	5	6	1916—W.D.
6	Beyer, Peacock (2467)	1885	E. & W.J.	2–4–0T	17 × 24	5	6	1916—W.D.
1	Yorkshire Eng. Co.		P.S.N.W.R.	2–4–0T	15 × 22	5	0	1895—Cannock & R. Colliery
7	Crewe (652)	1863	L.N.W.R.	0–6–0	17 × 24	5	0	W 1920
8	Crewe (657)	1863	L.N.W.R.	0–6–0	17 × 24	5	0	S by 1911
9	Crewe (894)	1866	L.N.W.R.	0–6–0	17 × 24	5	0	S 1903
10	Beyer, Peacock (3613)	1895	E. & W.J.	0–6–0	17 × 24	5	1½	L.M.S. 2304; W 1924
11	Beyer, Peacock (3812)	1896	E. & W.J.	0–6–0	17 × 24	5	1½	L.M.S. 2305/2398; W 1930
12	Beyer, Peacock (4126)	1900	E. & W.J.	0–6–0	17 × 24	5	1½	L.M.S. 2306/2399; W 1930
13	Beyer, Peacock (4495)	1903	E. & W.J.	2–4–0	17 × 24	6	1	L.M.S. 290; W 1924
14	Beyer, Peacock (4496)	1903	E. & W.J.	0–6–0	17 × 24	4	9	L.M.S. 2307; W 1926
15	Beyer, Peacock (4633)	1904	E. & W.J.	0–6–0	17 × 24	4	9	L.M.S. 2308; W 1924
16	Beyer, Peacock (4735)	1906	E. & W.J.	0–6–0	17 × 24	4	9	L.M.S. 2309; W 1927
17	Beyer, Peacock (5102)	1908	E. & W.J.	0–6–0	18 × 24	4	9	L.M.S. 2310; W 1925
18	Beyer, Peacock (5103)	1908	E. & W.J.	0–6–0	18 × 24	4	9	L.M.S. 2311; W 1927
7	Brighton	1884	L.B.S.C.R.	0–6–0	18¼ × 26	5	0	L.M.S. 2303; W 1924

reproduced from OL10 *The Stratford-upon-Avon & Midland Junction Railway*

COMPONENT COMPANIES
of the Stratford-upon-Avon & Midland Junction Railway

East & West Junction Railway.

An Act, passed on 23rd June 1864, authorised the construction of the above, from the Northampton & Banbury Junction Railway at Green's Norton, to a junction with the Stratford and Honeybourne branch of the GWR at Old Stratford, a distance of 33 ¼ miles. Running powers were granted over the Northampton & Banbury Junction Railway from Green's Norton Junction to Blisworth on the LNWR main line.

Financial difficulties, however, postponed commencement of actual construction, and the first section, from Fenny Compton to Kineton (6 ¼ miles), was opened to traffic on the 1st June, 1871. It was single line throughout, with a siding connection with the GWR at Fenny Compton.

The other sections of the line – Kineton to Stratford (9 ¼ miles), and Fenny Compton to Green's Norton Junction (17 ¼ miles) – were both completed and opened for traffic on the 1st July. 1873.

Evesham, Redditch & Stratford-upon-Avon Junction Railway.

An Act, passed on the 5th August, 1873, authorised its construction, from Stratford to Broom Junction (7¾ miles), on the Evesham & Redditch Railway. Running powers were granted between Evesham and Redditch, and provided for the line to be worked and managed by the E&WJR. It was opened for traffic on the 2nd June, 1879.

Easton Neston Mineral & Towcester Roade & Olney Junction Railway

Promoted by the E&WJR, the Act authorising this line was passed on 15th August 1879, and provided for a railway 10 ½ miles long, connecting the E&WJR at Towcester with the LNWR at Roade, and the Bedford & Northampton Railway at Ravenstone Wood Junction. The latter railway, opened on 10th June 1872, was absorbed by the Midland in 1885.

Financial and other difficulties prevented work from commencing and it was not until 13th April 1891 that this line was open to traffic, and for goods only. In 1882 the line was authorised to change its name to the Stratford-upon-Avon, Towcester & Midland Junction Railway.

Northampton & Banbury Junction Railway

Parliamentary powers for the above were obtained on 9th July 1847, and it was originally intended as a route from Northamptonshire to South Wales. The Act provided for a line from a junction with the Northampton & Peterborough Railway near Gayton Wharf. crossing the LNWR main line at Blisworth and passing through Towcester to a junction with the Buckinghamshire Railway at Cockley Brake, near Farthinghoe.

The project lapsed, however, and on the 28th July 1863, a fresh Act was obtained for a railway of the same name. This was to cover the same ground, but with its eastern terminus at Blisworth. Running powers were granted between Blisworth & Northampton over the LNWR – never exercised – and between Cockley Brake Junction and Banbury.

A further Act was obtained in 1865 for an extension to Chipping Norton and Blockley in the direction of Gloucester, and in 1866 a further Act, authorising an extension from Blockley to Ross, with running powers over the Midland from Beckford to Tewkesbury, over the Ross and Monmouth line, and over the Worcester, Dean Forest & Monmouth. The name of the railway was obviously quite inadequate for these grand schemes, and the 1866 Act changed it to the The Midland Counties & South Wales Railway. By 1870 it was clear that most of the 96½ miles of authorised railway would remain visionary, and the name was changed back to 'Northampton & Banbury Junction'. Even so, it did not reach either place with its own metals.

The line was built by Aird & Son. The section between Blisworth, where the N&BJR had a station of its own alongside that of the LNWR, and Towcester, was opened in May 1866. The remainder, from Towcester to Cockley Brake Junction, was not opened until 1st June 1872, though goods traffic had been carried as far as Helmdon since August 1871.

The railway led an uneventful existence, worked after 1st November 1876 by the LNWR until it was merged in the SMJR.

Due to the involved affairs of the three first companies above mentioned, and their financial insolvency (see OL10 *The Stratford-upon-Avon & Midland Junction Railway* by J. M. Dunn), a Bill was presented to Parliament in November 1907 for the amalgamation of the Evesham, Redditch & Stratford-upon-Avon, the East & West Junction, and the Stratford-upon-Avon, Towcester & Midland Junction Railways, and the rearrangement of the capital.

Thus the Stratford-upon-Avon & Midland Junction Railway came into being, commencing operations as such on 1st January 1909. In the same month, negotiations began for the purchase of the Northampton A Banbury Junction Railway. On the 29th April 1910 the acquisition received Royal Assent.

The Easton Neston Mineral & Towcester, Roade & Olney Junction Railway line where it leaves the main Midland line at Ravenstone Wood Junction **(844 532)** looking south past the signal box.

Easton Neston Mineral & Towcester, Roade & Olney Junction Railway

This unit of the Stratford-upon-Avon Junction Railway is still in partial use, though only for the purpose of storing obsolete vehicles, or broken rolling stock which is awaiting repair.

The line was cut during construction of the M1 Motorway, and has been used for storage only ever since although the line was restored to running order when the M1 overbridge was completed.

The bridge carrying the track over the LNWR line at Roade has now been removed, thus cutting off the section through to Towcester. The latter part is unused for any purpose.

The signal wiring has all been removed and taking up of the track will probably not be long delayed.

The line was planned and excavated as a double track, as can be seen from the position of the overbridges. No double track, however, was ever laid, the line being single throughout its history. It is also worthy of note that there is no tunnel on the line, nor anywhere else on the whole SMJR system.

Ravenstone Wood Junction.

A view of the line leading away from Ravenstone Wood Junction in the direction of Quinton and Hartwell.

Horton overbridge **(818 535)** on the A40. Note the position of the track, allowing for a second line of metals.

Quinton – Hartwell bridge **(784 527)** with stored rolling stock.

Quinton – Hartwell bridge, looking towards the M1 bridge. The road was cut at this point through the SMJR track, and the latter was closed for nine months until this bridge was completed.

View of what remains of Salcey Forest Station **(814 536)**. Up to ten years ago, the station building was intact, and used to house track staff.

View of the only building still standing in August 1961.

The old entrance to Salcey Forest Station, which can only be reached across fields.

A line of wagons standing at the remains of the platform. These vehicles are for the main part stored awaiting repair, though some, according to instructions attached, are merely idle and unwanted.

Two photographs (*above and facing page*) taken from the overbridge **(803 538)** which carries a bridle road from Piddington to part of Salcey Forest. It can be seen how overgrown this section is, although the track seems to be in reasonable repair.

This photograph is of the line between **(795 535)** the overbridge for the bridle road from Piddington to part of Salcey Forest, and the one carrying the road from Quinton to Salcey Forest and Hartwell. Here again the track is in reasonable condition, but the embanked parts are eroding away. This was particularly noticeable near Salcey Forest Station.

View from Roade – Hartwell overbridge **(766 514)**, looking in the direction of Roade.

A view showing where the Easton Neston Mineral & Towcester, Roade & Olney Junction Railway line crossed over the main LNWR track at Roade, only the abutments of the overbridge remain – in the middle distance of this photograph.

Easton Neston Mineral & Towcester, Roade & Olney Junction Railway

A photograph taken from the Easton Neston Mineral & Towcester, Roade & Olney Junction Railway line shows that the line is completely severed at this point, with the removal of the former iron bridge, all that remains are the bridge abutments. The line of trees running right on the opposite side of the LNWR line marks the course of the spur that connected the two railways.

The bridge abutments.

Easton Neston Mineral & Towcester, Roade & Olney Junction Railway

The bridge abutments, with the main LNWR line passing in between.

View looking away from the bridge abutments, towards Quinton. with Roade – Ashton overbridge in the background.

Stoke Bruerne Station **(735 505)** had a short operating life, from 1st December 1892 until 30th March 1893, when regular passenger trains from Olney to Towcester ceased operation.

Shutlanger road overbridge **(724 504)**.

The line approaching its end at Towcester. There were no signal boxes on the section, as it was worked as one block the absence of any signal wiring is conspicuous.

The junction **(691 498)** of the Easton Neston Mineral & Towcester, Roade & Olney Junction Railway with the Northampton & Banbury Junction Railway at Towcester. The section from Ravenstone Wood to the junction at Towcester was 10½ miles long.

Northampton & Banbury Junction Railway

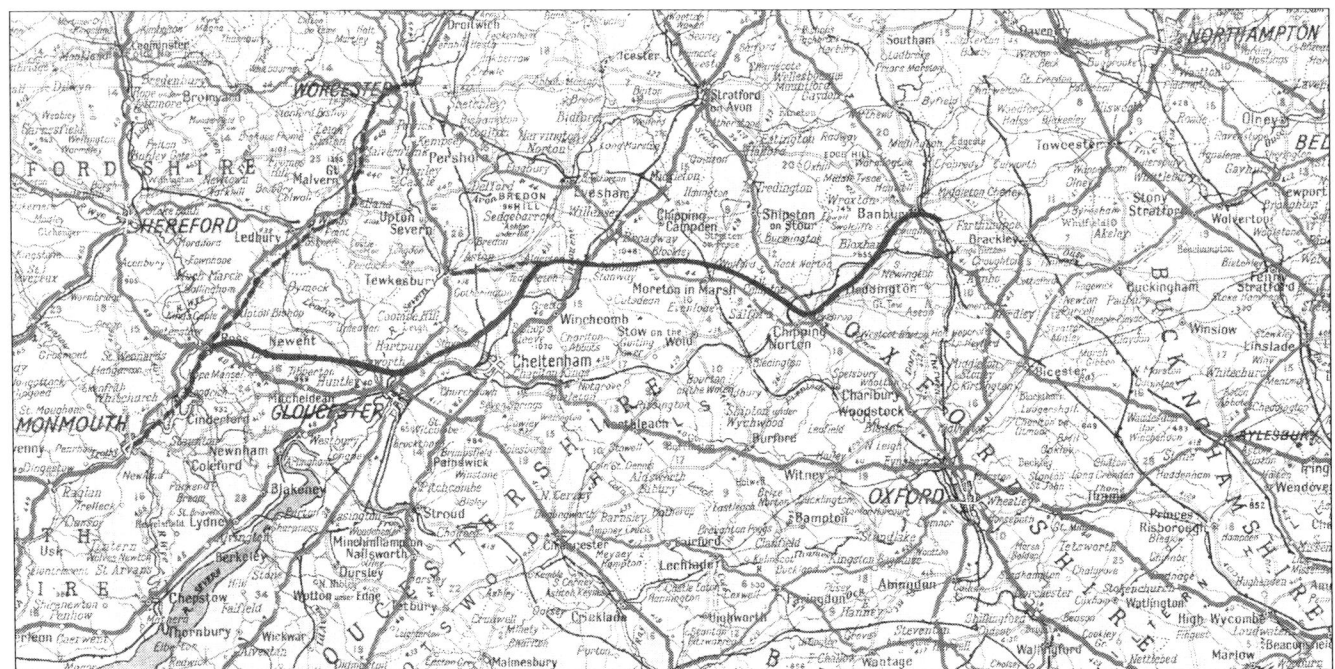

The above map is intended to give only a general idea of the proposed extension to the above railway, which was to have been known by the title The Midland Counties & South Wales Railway.

Its proposed own rails are shown in solid black, while the agreed running powers are shown as dotted lines.

Running powers from Beckford to Tewkesbury were to be over Midland metals, while the section from Worcester to Monmouth was to be over the Worcester, Dean Forest & Monmouth. This line remained a dream, and the grandiose title was dropped and changed back to the Northampton & Banbury Junction Railway.

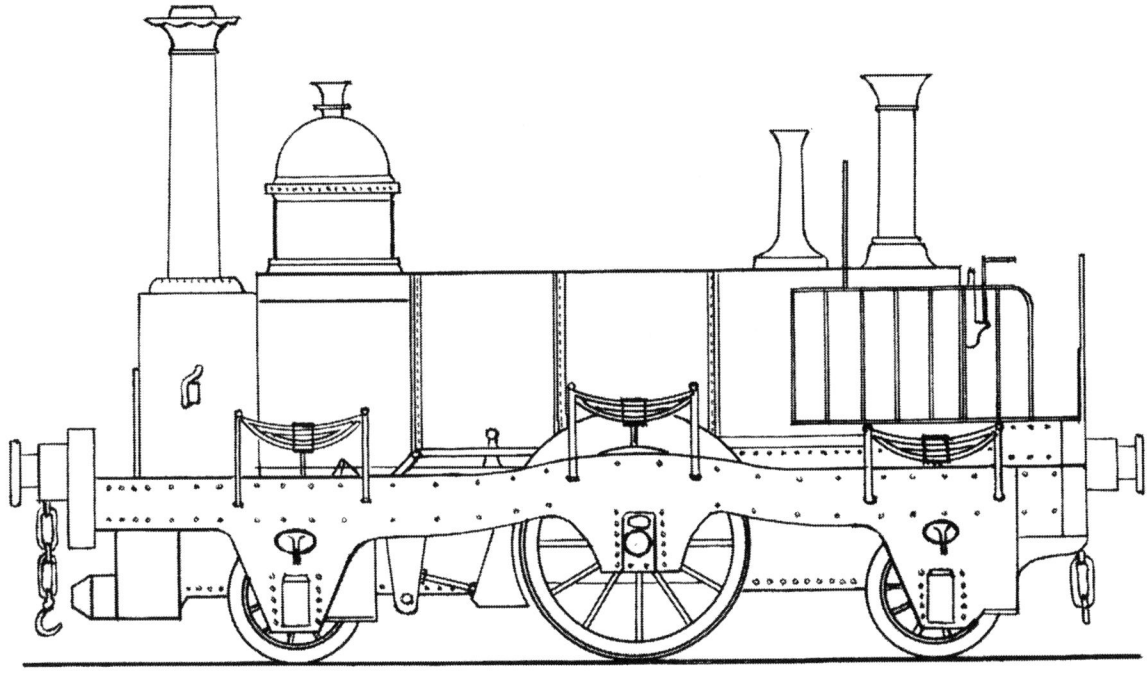

The Northampton & Banbury Junction Railway, though it brought no engines with it to the SMJR in 1910, had had its own locomotive adventures. To work the line when opened in May 1866, two 0-4-2T, and two 0-4-2 tender engines were ordered, but the company could not pay for them, so they were not delivered. Engines had to be hired, therefore, and LNWR No. 1125, a 2-2-2 *Sharpie*, was one. It was taken to Blisworth on 1st March 1866. By 29th March it had developed a cracked cylinder, and it had left the N&BJR before July.

Northampton & Banbury Junction Railway

This section of the Stratford-upon-Avon & Midland Junction Railway has only one part still in use, between Blisworth and Towcester, and that only for the passage of infrequent goods traffic.

The line between the junction with the East & West Junction line near Green's Norton, and Cockley Brake Junction with the old Buckinghamshire line, is now entirely derelict, the track having been taken up some years ago.

Wappenham Station and sidings are non existent save traces of foundations and sidings trackbed, but Helmdon Station and other buildings are still standing, though mainly derelict.

Some small business seems to be being carried on in the goods buildings, and the station forecourt has become a bus park.

The signal box in the photograph below is that belonging to the SMJR. At one time the SMJR had its own station at Blisworth, alongside that of the LNWR.

Blisworth Junction **(719 548)** facing south, with the main LNWR Line in the background. The double track is beginning to converge into the single line.

Blisworth Junction facing south.

The line leading away from the junction at Blisworth.

View from the Gayton- Blisworth overbridge **(709 541)**, towards the junction.

Gayton Blisworth overbridge, south west towards Towcester.

Gayton Wilds overbridge **(705 531)** south west towards Towcester

Iron bridge **(692 502)** over farm entrance near Towcester. It was cast by Barwell & Co., iron founders, Northampton in 1864.

The line approaching the junction **(692 502)** with the Easton Neston Mineral & Towcester, Roade & Olney Junction Railway at Towcester

Towcester Station **(688 494)**, from thr south west end of bridge over the A5.

Looking west from the bridge over A5 at Towcester.

A view of Green's Norton Junction **(674 488)**, in a westerly direction, with the East & West Junction Railway metals going to the right, the trackbed of the Northampton & Banbury Junction Railway to the left of the plate-layers' hut.

A view of the junction, taken looking along the Northampton & Banbury Junction Railway trackbed. The actual point of the junction would appear to be about halfway between the first and second telegraph poles.

This picture was taken in an easterly direction, towards Towcester, at approximately the point of juncture.

Looking north east, towards Green's Norton Junction with East & West Junction Railway from the overbridge on the road from Abthorpe to Bradden **(651 472)**.

Looking south west, towards Wappenham Station from beneath the Abthorpe to Bradden road overbridge.

Slapton – Abthorpe overbridge **(639 465)**, looking north east.

Similar view to that above, but with more of the trackbed shown.

Looking along trackbed from the Slapton – Abthorpe overbridge position. Remains of the Wappenham sidings can be seen to the left.

Photograph of the remains of Wappenham Station **(635 465)**.

This picture and the previous photograph (page 35 lower) were taken at about 7.45 in the morning, a cloudy & windy one. In August 1961 I [Stephen Thompson] walked up and down the platform. Imagining that I was waiting for an early morning train. The effect was eerie, and I should have been little surprised if *Sharpie* had come puffing from under the Slapton overbridge, with some antique 4-wheel stock behind her.

Passenger services were withdrawn from both the above station and the one at Helmdon on 2nd July 1951. Goods trains continued to serve these stations until 29th October 1951.

Entrance to Wappenham Station and sidings. From Abthorpe Wappenham road.

Trackbed looking south west from Wappenham – Weedon Lois overbridge **(624 464)**.

North east view of trackbed from Helmdon – Wappenham overbridge **(606 445)**.

Helmdon Station **(588 439)**, looking to the west.

Helmdon Station from the forecourt.

Trackbed heading west from Helmdon. The viaduct **(584 436)** over the Northampton & Banbury Junction Railway carrying the old Great Central line is in the background.

Facing west, from the Crowfield – Greatworth overbridge **(563 430)**.

Bridge over the Banbury Middleton Cheney – Brackley road **(549 415)**.

The end of the line. The junction at Cockley Brake **(546 414)** with the old Buckinghamshire line, later the LNWR. The picture is of the LNWR line, with the Northampton & Banbury Junction Railway line joining at the apex of the curve in the photograph.

The actual point of junction. The Northampton & Banbury Junction Railway line approaches at the set of fencing, which now seals off the old trackbed.

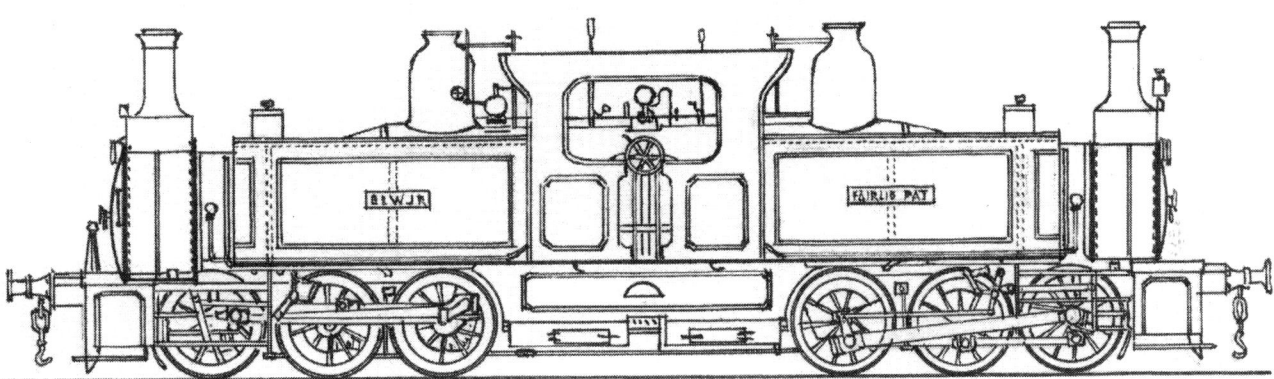

This 0-6-6-0 locomotive, No. 1 of two Fairlies purchased by the E&WJR, from the Yorkshire Engine Co. in 1876. Originally designed and built for a Mexican railway, both this engine and the other Fairlie, a single boiler 0-4-4T, named *Robert Fairlie*, ran on the E&WJR for about a year, and then, after passenger traffic was suspended in 1877, were disposed of in 1878.

The East & West Junction Railway

It is agreeable to know that this line appears to be in practical use once again, and not in danger of suffering the fate of the rest of the SMJR lines.

On 24th April 1960 the new signalling arrangements at Stratford-on-Avon were brought into operation, and the new connections between. the former SMJR line, and the Banbury – Birmingham and the Cheltenham – Honeybourne – Birmingham main lines of the GWR were ready for full use.

The link at Fenny Compton was ready some weeks earlier, on 7th March 1960. Here the SMJR line approaching from the east bridges the Banbury-Birmingham line and descends to run parallel with it for a short distance before resuming its westerly path; a flat junction between the two routes, controlled by a new signal box, was thus possible. At Stratford, however, the SMJR bridges the Cheltenham – Honeybourne – Birmingham route at right-angles, and a new double track spur had to be laid from Stratford (Old Town) Station, on the SMJR to a trailing junction with the line to Honeybourne near the town's Racecourse Station. A new signal box at Evesham Road Level Crossing, on the Honeybourne line, is in command of the new Stratford link, and takes over the work of the cabins formerly at Evesham Road, SMJR Junction and Old Town, though a ground frame is to be maintained for shunting operations at the SMJR goods station.

From 24th April a great deal of traffic from Woodford Halse and Banbury to South Wales has been re-routed via the new connections to sidings established at Honeybourne, where engine or crew changing is effected. Between Stratford and Fenny Compton the SMJR line has been improved and the disused station platforms have been trimmed to allow ex-GWR. engines to work over it, but it is expected that most of the traffic will be operated by Class 9F 2-10-0s, and War Department "Austerity" 2-8-0s. The completion of this new route between Banbury and South Wales will enable the Stratford – Broom Junction line to be closed completely; its two intermediate stations were shut down early in April 1960.

Later, the new working arrangements over the SMJR line were postponed, because Banbury engine men desired more time to learn the road from Fenny Compton to Honeybourne; full use, therefore, of the new link between the SMJR and the W. R, main line at Stratford was deferred until 12th June 1960.

Goods working ceased over the Stratford – Broom Junction section on 30th June, 1960.

The line looking east from a point **(669 488)** a quarter of a mile from Green's Norton Junction.

The line approximately from the same point as the previous picture, westwards.

Looking east towards Green's Norton Junction from the Green's Norton Bradden overbridge at Kingthorn Wood **(659 493)**.

Looking west from Kingthorn overbridge in the direction of Blakesley

Blakesley Station **(625 499)**, from the overbridge, looking east.

From Blakesley overbridge looking west.

Looking east, from the overbridge between Canons Ashby and Moreton Pinkney.

Moreton Pinkney Station **(576 499)**, from the overbridge. The station is entirely derelict, although the track is in good condition

Looking west towards Woodford Halse from the Eydon Canons Ashby overbridge **(557 504)**.

Eydon Woodfordhill bridge **(551512)**, of cast iron with brick abutments, the type generally used over narrow roads, farm entrances, etc.

Near Woodford Halse (542 516) looking west from the overbridge that crosses the E&WJR and southbound spur to the Great Central Railway. The train is heading towards the bridge that crosses the GCR line.

Stephen Thompson's Notebook, 1961

Looking west from the Woodford Halse Junction overbridge along the former southbound spur to the Great Central Railway. The line of the northbound spur can be made out behind the building on the right. The white signal box that controlled both spurs is in the distance.

Looking east from the overbridge the spur has been cut before it joins the Great Central line leaving a set of sidings.

The bridge carrying the track over the Hinton – West Farndon road **(531 522)** from the north.

The track over the bridge pictured above. Looking eastwards in the direction of Woodford Halse.

The line approaching Byfield Station **(519 528)** from the east. Taken from the Byfield – West Farndon overbridge. The line here is double tracked, though this of course merely constitutes passing points, as in most of the stations.

Byfield Station, taken from the overbridge mentioned on the previous page. Here again the double track may be seen, and also the good condition of this part of the line.

Two views from the Aston-le-Walls – Lower Boddington overbridge **(495 514)**, the upper one facing east in the direction of Byfield, the lower one west towards Fenny Compton.

This picture to the right shows the very beautiful course of this railway, running as it does through some of the most unspoilt countryside in the Midlands. The line can be seen to the right of the second tree from the left of the picture.

Views from the Claydon – Claydon Hay overbridge **(461 512)**, in the easterly direction of Aston-le-Walls.

THE EAST & WEST JUNCTION RAILWAY

View west towards Fenny Compton from the Claydon – Claydon Hay overbridge.

A farm bridge **(444 515)** carrying the line in the region between Claydon and the junction of the Claydon road and A423.

A picture of the main GWR. line, from the overbridge **(449 513)** between the A423 and Claydon. The bridge carrying the E&WJR over the GWR line is in the background. The end of the train is just clear of this bridge.

The section ¾ of a mile before Fenny Compton Station, where the E&WJR line runs alongside that of the GWR to Birmingham, taken from A423 overbridge **(436 524)**.

View of the derelict Fenny Compton E&WJR Station **(428 528)**, taken across the metals of the GWR line to Birmingham.

The main GWR line at Fenny Compton, looking towards Birmingham, with the old GWR station and new signal box. The E&WJR line can be seen curving away to the left of the new signal box.

The track looking westwards from the Northend – Knightcote overbridge **(400 534)**, in the direction of Burton Dassett Halt.

Burton Dassett Halt **(378 526)**, or what was a halt once. This is the point of junction of the ill-fated Edge Hill Light Railway and the E&WJR. As far as one can judge, it approached from Edge Hill, from the left hand side of the picture, now obliterated by the army installation visible on the left.

Burton Dassett Halt facing east. The actual building is still there, just out of the left hand edge of the photograph.

The Edge Hill Light Railway 1920 – 1925

The site of the ironstone quarry workings on Edge Hill. The picture to the left shows the remains of the foundations of the engine house which operated the cable section of the railway. The picture above shows the overbridge carrying the B4036 along Edge Hill, with the engine house remains in the foreground. The photograph is taken looking along the track of the line down to Burton Dassett.

The trackbed descending Edge Hill, taken from the overbridge shown on the previous page.

This ill-fated railway was intended to transport ironstone found in the hills, when, due to conditions brought about by the War of 1914-1918, such working became economic. The picture shows the bridge carrying the line over the Knowle End – Arlescote road **(483 486)**.

THE EDGE HILL LIGHT RAILWAY 1920 – 1925

The type of rail bed used by the Edge Hill Light Railway on this very steep gradient. Steel cables controlling the trucks can be found still.

View looking down the track way over the valley towards Burton Dassett the terminus of the line.

Back to the E&WJR two views of the line taken from the Gaydon – Kineton overbridge **(338 515)**, the upper one to the east, the lower to the west as it approaches Kineton Station.

The entrance to Kineton Station **(330 512)** from the Wellesbourne road, with the station itself in the background.

The station building across the tracks.

Kineton Station, which is unused. The platform has been cut back and all the brickwork has gone, leaving a sloped earthwork.

Looking east in the direction of Kineton.

Facing west towards Ettington Station. These photographs are taken from the bridge **(293 508)** carrying the line over the Foss Way, on the section of road that connects the B4086 and the A422. The track here is in excellent condition, with good type ballast. The bridge itself is over the usual brick buttresses, and of girder construction, like the remainder of the bridges over narrow roads, farm entrances and so forth.

Looking east towards Kineton.

These views were taken on the bridge over the road from Elder Tree Copse to Walton **(284 506)**. The track and bed are excellent on this section, and note the use of concrete ties in this locality. As the wooden sleepers require replacement, so concrete ones take their place. It was also to be noticed on part of the track in Byfield Station. This bridge is of brick construction throughout.

Ettington Station **(270 503)**, looking east. It is unused. The platform has been cut back, to allow former Great Western Railway engines to work the line.

Picture taken at Ettington Station looking west from the Ettington – Warwick overbridge (A429) in the direction of Goldicott Cutting.

Goldicott Cutting **(247 506)** in the westwards direction. The Banbury – Stratford main road, (A422) crosses the cutting by the overbridge visible in the photograph.

Goldicott Cutting is some 60 feet deep, and 600 yards long. It is cut through limestone, and has given a good deal of trouble in its time due to shifting. It is excavated to take only a single track in spite of the fact that all bridges on the Towcester – Stratford section were built for double line. These photographs were taken from the A422 overbridge.

Goldicott Cutting in the eastwards direction, towards Ettington Station

The picture to the above shows the line approaching Stratford from the east, with Clifford sidings **(266 538)** to the right. The track here is double. this having been carried out in 1943, as far as Stratford.

Clifford sidings, with passenger rolling stock stored upon it.

The line approaching Stratford-upon-Avon Station from the east, over the bridge spanning the Avon, a glimpse of which can be seen in the bottom left hand corner of the picture.

Looking west through the station **(199 540)** which is unused today, save for a certain amount of shunting. The weighbridge is in use by various local industries.

The water column in the unusually wide space between the tracks.

A view of the up line platform, towards the start of the Evesham, Redditch & Stratford-upon-Avon Junction Railway.

A signal box on the up platform which appears to have been disused since about 1913. The one replacing it, which stood at the end of the same platform, has been entirely removed.

The only vestige of the signalling arrangements remaining is the ground frame for shunting.

The entrance to Stratford-upon-Avon Station, from the extension to Trinity Street. The weighbridge that is still in use can just be seen on the left.

The station buildings from the forecourt.

View of the of the goods yard. The site of one the signal boxes was just to the left of the wagon facing end on, and to the right of the telephone pole. Nothing remains of this box, all signalling now done at this station being carried out by a ground frame.

The new spur on the left, the former Evesham, Redditch & Stratford-upon-Avon Junction Railway. line in the centre, and old north spur on the right.

The old north spur to Stratford Station GWR.

The new spur, for South Wales coal traffic, going out left centre to a trailing junction with the line to Honeybourne, while the disused Evesham, Redditch & Stratford-upon-Avon Junction Railway is in the foreground, crossing the Cheltenham-Birmingham line by a bridge.

Stratford-upon-Avon Racecourse Platform looking south along the Honeybourne line. The new spur to the Stratford-upon-Avon and Midland Junction joining just past the end of the platform.

A Class 5 4-6-0 passes Racecourse Station.

The New Evesham Road Signal Box, in control of the new Stratford link.

The Stratford & Moreton Tramway

The reason for the inclusion of this moribund system in a record of the SMJR is mainly because it makes a fleeting contact with the SMJR near Clifford Sidings, but also for its general historical interest concerning railways in this locality.

History

In 1820 William James drew up plans for a Central Junction Tramway, to run from the basins of the Stratford Canal at Stratford-upon-Avon, through Moreton-in-Marsh, Oxford, Thame and Uxbridge to London. A line was to come from the coalfield north of Coventry, through Shipston, to join the main line between there and Moreton.

Due to Lord Redesdale, and developed from the above scheme a purely local plan was forthcoming, namely the Stratford to Moreton Tramway, authorised by an Act of 28th May, 1821. Authorisation was for a tramway from the canal basin at Stratford to Moreton, with a branch to Shipston from Blackwell Bushes, nine miles from Stratford.

Financial trouble held up development, but the line from Stratford to Moreton was opened for horse-drawn traffic on 5th September 1826.

In 1836, a 2½ mile branch to Shipston was opened from the 9½ milepost from Stratford. The tramway company supplied no transport, users having to provide their own wagons, and pay tolls on the goods carried.

On 1st May 1847 the Oxford, Worcester & Wolverhampton Railway (OWWR) took a perpetual lease of the Tramway, at a rental of £175 per annum. In 1853 the Tramway was closed for improvements for a period of six months, when the gauge was altered to standard from the original 4 ft. On 1st August 1853 Mr. Bull of the George Hotel, Shipston, commenced a twice daily passenger service over the entire system. A railway carriage adapted for horse haulage was employed.

On 11th July 1859 a branch from Honeybourne to Stratford, was opened by the OWWR, which largely rendered the tramway to Stratford redundant. In 1860, the OWWR merged with other companies to form the West Midland Railway, and as a result of amalgamation of 1863 the lease passed to the GWR. Attempts to close the tramway were successfully resisted.

In 1881 traffic over the whole system was moderate, but the section from Ilmington to Stratford was described as practically closed. Traffic was worked by horse, both by the GWR, and by local traders, who continued to pay tolls.

In 1882 Parliamentary approval was obtained for a steam-worked branch from Moreton to Shipston, using as much as possible of the tramway. Difficulties arose over the use of steam on the existing lines where roads were crossed, as no bridges had been built, the Act of 1833 demanding such bridges if steam were used. Revocation of the 1833 Act was obtained on 1st July 1889 and the line was opened with four mixed trains in each direction.

The Stratford – Longdon section was last used in 1900, but the track was not lifted until 1917. The line was abandoned by Act on 4th August 1926. In 1917 major war economy cuts took place, and the service became two trains per day each way. Stretton Station was temporarily closed. After the First World War, services never exceeded three passenger or mixed trains per day, on weekdays only. In 1929 passenger services ceased entirely, on 8th July, GWR bus

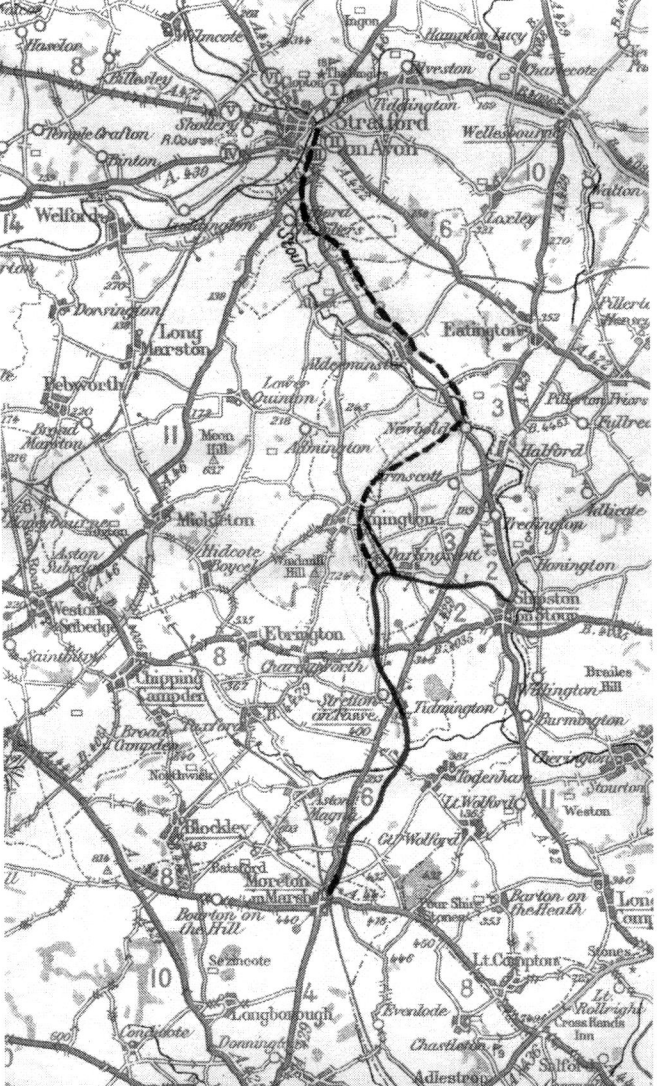

services being substituted. A freight service has continued to operate, though the two intermediate stations were closed in the last war.

Locomotives used have been Dean 0-6-0s, replaced by BR standard 2-6-0s, and occasionally by 1600 class 0-6-0 pannier tanks.

Platforms at Stretton-on-Fosse and Longdon Road remain, though the buildings have been removed. At Shipston, the buildings are intact.

The course of the old tramway can be seen going in the direction of Stratford, near Milepost 98, and at Longdon Road. At the point where the original branch to Shipston joined the main line of tramway, standing in the "V" formed by the two lines, is a house known as Junction House, built by the tramway company to control traffic. A small cottage, standing near the line at Todenham Lane Crossing, is also of tramway origin. Other original buildings may be seen in the station yard at Moreton. A familiar feature is the footbridge near the Clopton Bridge at Stratford, by which the tramway was carried over the River Avon. Other buildings are at Newbold and Ilmington.

The Stratford & Moreton Tramway

To the right are three views of the old wagon standing on Old Tramway Walk, Stratford-upon-Avon, and above is shown the descriptive plate attached to this wagon. It appears that the place at which this wagon stands must be the approximate point of the start of the tramway, for the original plan for the tramway provided for the line to run to Moreton from the canal basin, Stratford. This basin appears to be the stretch of water to the left of the town entrance from the Clopton Bridge. A derelict lock lies between it and the river.

The tramway bridge over the SMJR line at Clifford Sidings.

The bridge that carried the tramway over the Avon at Stratford.

Evesham, Redditch & Stratford-upon-Avon Junction Railway

This part of the SMJR system is now defunct, with the bringing into operation of the new spur referred to previously. It has been closed since 30th June 1960 to all traffic.

A very steep gradient carries the line from Stratford Station to the bridge over the Birmingham- Honeybourne – Cheltenham metals, so steep in fact that frequently engines were unable to surmount it at the first attempt, and had to back down into the station to assault it a second time.

This section is single line throughout the 7¾ miles to Broom Junction.

It is almost certain that this part of the SMJR will suffer the same fate as the N&BJR, although to date the metals remain in position. All the stations are closed and Binton Station has been disposed of to an Evesham traction engine firm.

The line leaving Stratford Station **(199 540)** towards the bridge over the main line. The new spur is on the left.

The bridge **(196 542)** carrying the line over the main Birmingham – Honeybourne – Cheltenham line as the ER&SJR leaves Stratford. The view is taken looking west to Binton.

Looking back along the cutting towards Stratford, from the overbridge **(163 528)** on the Luddington A439 road. A newly laid out orchard lies behind the railings to the right.

Two pictures from the A439 overbridge, looking in a westerly direction towards Binton. The typical Vale of Evesham scenery begins to be evident around this point. Sandfields Farm buildings are in the centre of the top image.

Evesham, Redditch & Stratford-upon-Avon Junction Railway

Two views of the line where the Luddington A439 road runs parallel with the track **(157 529)**, for about ¾ of a mile. The upper photograph faces west. The lower view looks back east, in the direction of Stratford.

In this picture of the Welford bridge **(145 533)**, the A439 can be seen in the background. Also to be noted is the comparatively new brickwork of part of the bridge, and new girder. Flooding in all likelihood caused undermining of the foundations, making extensive repairs necessary. The depth of possible flooding can be seen from the gauge in the foreground.

The line approaching the station for Binton, the photo being taken from the platform, looking towards Stratford.

The entrance to Binton Station **(141 532)** from the A439.

View of the platform and station buildings. Binton Station is now owned by a firm dealing with traction engines, whose offices are housed in the building in the foreground.

This view is taken from the bridge visible in the lower photograph on page 90, looking back towards the station.

Wasen Hill overbridge **(138 531)**, looking west

Two pictures from A439 over bridge **(124 529)** near Cranhill, the top one looking east, lower same viewpoint looking west.

Bidford – Temple Grafton overbridge **(106 528)** upper view facing east, lower view facing west. It can readily be appreciated from these pictures the pleasant nature of the country through which the line runs.

The line at Bidford-on-Avon Station, looking in the direction of Binton with the points leading off right to the goods yard. On the right is the Bidford Brickworks.

The loading gauge, with the weighbridge office in the background.

The overbridge of Bidford-on-Avon, looking in the Broom direction, showing concrete ties.

The overbridge again, with the platform in foreground.

Evesham, Redditch & Stratford-upon-Avon Junction Railway

Details of the overbridge.